André Citroën's answer to the Ford Model T, the simple Type A powered by a 1.3 litre side-valve engine. Introduced in 1919, it was built until 1921.

T0174468

THE CITROËN

Jonathan Wood

Shire Publications Ltd

CONTENTS

Printed and bound in Great Britain.

Published by Shire Publications Ltd, Midland House, West Way, Botley, Oxford OX2 0PH, UK. Copyright © 1993 and 2003 by Jonathan Wood. First published 1993. Second edition 2003. Transferred to digital print on demand 2011. Shire Library 289. ISBN 978 0 74780 563 2. All rights reserved. No part of this publication may be reproduced or transmitted in any form or by any means, electronic or mechanical, including photocopy, recording, or any information storage and retrieval system, without permission in writing from the publishers.

MIX
Paper from
responsible sources
FSC
www.fsc.org   FSC® C013604

British Library Cataloguing in Publication Data: Wood, Jonathan. Citroën. — (Shire Albums Series; No. 289). I. Title. II. Series. 629.222. ISBN-10 0-7478-0563-6.

Editorial Consultant: Michael E. Ware, former Director of the National Motor Museum, Beaulieu.

ACKNOWLEDGEMENTS
The author gratefully acknowledges the assistance of Citroën UK Ltd and Automobiles Citroën in supplying photographs for this book.

Cover: *A 1973 DS23 of 2.3 litres with carburetor or electronic fuel injection and available with manual, 'hydraulic' or automatic transmission. This is the more expensive Pallas version, instantly identifiable by its stainless steel rubbing strips. (Courtesy of John Reynolds)*

Below: *The famous Citroën 'trèfle' (clover-leaf) three-seater body. It graces a 5CV chassis, which first appeared in 1921, was revised in this longer-wheelbase CS form in 1925 and remained in production until the following year.*

André Citroën (1878-1935), the Polytechnicien who became France's largest car manufacturer.

FROM GEARS TO CARS

A car company which has produced what has been described as the ugliest car of its generation as well as some of the most advanced, stylish and aerodynamically efficient automobiles in Europe, years before it was fashionable to do so, is clearly unusual. Yet the French company Citroën, which in 1934 successfully pioneered front-wheel drive, has been doing precisely that ever since. But, as will emerge, when the firm began building cars in 1919 there was little indication that such a sophisticated design philosophy lay ahead.

For the first sixteen years of this motor manufacturer's existence the firm was run by its founder, André Citroën. Born in Paris in 1878, he was not French but of Dutch-Jewish origins; his father, who was a diamond merchant, committed suicide when his son was two years old. The young Citroën did well at school and in 1894 he was fortunate to obtain a place at the exclusive École Polytechnique. On leaving that famous Paris institution in 1898, Citroën joined the French army and it is said that while on leave visiting relatives in Poland he spotted a pile of wooden double helical gear wheel stacked behind a workshop. He immediately recognised the mechanical efficiency of the V-shaped configuration and decided to leave the army and set up a business in Paris to manufacture such gears in metal. He did so in 1904 and one of his most celebrated commissions was for the steering gear of the ill-fated luxury liner *Titanic*.

3

Above: *The inverted Vs of the Citroën trademark are derived from the double helical gear business which André Citroën pursued from 1904, before he turned to car manufacture in 1919.*

Below: *Citroën's factory in the Quai de Javel, Paris, pictured in 1919. At this stage the assembly track was a manual rather than mechanised one. Here examples of the Type A, equipped with slave wheels, are under construction.*

Citroën's first contact with the motor industry came in 1908 when he was asked by the Paris-based Mors concern to streamline its production methods. This he did and output rose, to a limited extent, from about 280 in 1908 to around 800 vehicles a year in 1913, though the full effects of the *Polytechnicien's* work were cut short by the outbreak of the First World War in 1914.

The 36-year-old Citroën joined the French army and was made captain in the reserve. When he got to the front he found that France was suffering from an acute shortage of ammunition and he devised a plan to manufacture shells by introducing American mass-production methods. With the backing of the French armaments ministry, 12 hectares (30 acres) of market gardens at Quai de Javel, on the south bank of the river Seine in Paris, were acquired and a new factory was built there for armaments production.

The works was soon producing fifty thousand shells a day, an impressive total, but, when the war came to an end in 1918, Citroën was left with a fully equipped works, geared to quantity production but with nothing to build. He therefore decided to use this factory for car production and opted for the mass market in the same way in which Henry Ford had done with his famous Model T

The delightful Type C 5CV torpedo of 1921, powered by an 850 cc engine. This is a right-hand-drive example with the single door on the left-hand side. The model was initially available only in this two-seater form.

in the United States. That celebrated model had been so successful that, by 1920, it was estimated that one in every two cars in the world was a Ford.

Citroën had briefly contemplated a Panhard design during the war though the first car to leave the Quai de Javel works in 1919, named the Type A, was a simple, straightforward vehicle, geared to the mass-production process. It was powered by a 1.3 litre four-cylinder side-valve engine with three-speed gearbox; its suspension was quarter elliptic springs, to keep costs down, and it was fitted with distinctive disc wheels. For the badge, Citroën adopted the double helical gears of his earlier business and the twin chevrons have graced every Citroën car since then. In 1921 the B2, an improved version of the Type A, made its appearance and this 1.4 litre model was to remain in production until 1926.

A noteworthy development of the Types A and B2 was the half-track version, developed by M Kegresse, who had previously managed the garage of Tsar Nicholas II in Russia. A group of five B2 half-track Citroëns were the first cars to cross the Sahara desert, from Touggourt to Timbuktu; the journey of 1988 miles (3200 km) took three weeks, from 17th December 1922 to 6th January 1923.

All these cars were four-seaters but in 1921 Citroën announced a smaller model, the 850 cc 5CV, which was initially available only in two-seater form. It was essentially similar in design to the earlier cars though its engine had a detachable rather than a fixed cylinder head. Usually painted yellow, it soon acquired the nickname of *Petit Citron* (little lemon). Later came a longer-wheelbase version that permitted the fitment of a larger open body in which it was possible to seat three people. This so called clover-leaf design soon became the most popular variant in the range and production continued until 1926.

The 5CV was not a great performer: it was good for around 37 mph (60 km/h) but, once running, proved to be equally difficult to stop. However, it was ex-

5

André Citroën brought the American concept of the mass-produced pressed-steel saloon body to Europe in 1924. Here examples of the B12's saloon body receive attention at Quai de Javel in 1926.

tremely reliable and in 1925 a 5CV Citroën was chosen by Noel Westwood for his drive around Australia.

As the 1920s progressed, the reliable, straightforward Citroëns were seen in increasing numbers on the roads of France. By 1924 no fewer than 150,000 had been built and that year the Quai de Javel plant was producing cars at the impressive rate of 250 a day. This meant that the firm, along with the established Peugeot and Renault concerns, had become one of France's three largest motor manufacturers.

André Citroën also kept himself informed of technical innovations across the Atlantic and, in 1924, he introduced the concept of the mass-produced saloon to Europe. At this time cheap cars were mostly produced in open form and saloons were usually more expensive models as an open body uses less material than a closed one. The pressed-steel saloon body had been developed in the United States by the Budd company of Philadelphia and the first Citroën to be equipped with one

was the B2. However, its chassis was soon found to be unsuitable and the model was replaced for 1926 by the improved and more robust B12 powered by its predecessor's 1.4 litre engine and mostly built in closed form. Citroën's example was quickly emulated by Fiat in Italy and Morris Motors in Britain, and soon all the big motor manufacturers had followed suit. Yet another American invention, which Citroën built under licence from Chrysler, was its 'floating power' engine mountings that appeared on the French cars in 1931.

Citroën cars were soon being assembled abroad. A plant was established at Slough in Britain in 1926 and others followed in Belgium, Germany and Italy. Showrooms were opened in all the important European capitals and were particularly lavish. In London the Citroën premises in Piccadilly were a recreation of the tomb of Napoleon I at Les Invalides, Paris!

In addition to his undoubted success as a motor manufacturer, André Citroën displayed an impressive flair for publicity

6

The celebrated Citroën crossing of the Sahara in 1922-3 was followed by a more ambitious 12,428 mile (20,000 km) expedition from North to South Africa between October 1924 and March 1925. Here one of the eight B2 half-tracks on this so-called Black Cruise receives local help in the swamps of Nyasaland.

Citroën's Yellow Cruise followed in 1931. In April six C6 half-tracks left Peking to meet seven similarly equipped C4s starting from Beirut. They converged in Aksou in the Sinkiang province of China and then all travelled to Peking, arriving in April 1932. Here one of the half-tracks is being gingerly pulled over a rudimentary bridge.

The B14 was introduced for 1927. This flat-roofed saloon body was essentially similar to that of the B10 and B12 though the slightly longer chassis permitted the introduction of a small trunk at the rear.

Quai de Javel's first six-cylinder model, the AC6 (for André Citroën 6), was introduced at the 1928 Paris Motor Show. The 2.4 litre side-valve engine gave the model a top speed of around 62 mph (100 km/h). The wheel trims and bumpers of this 1929 car are contemporary embellishments.

André Citroën was a master of publicity and from 1925 the Eiffel Tower in Paris was illuminated with his company's name. It was an initiative which did not amuse his great rival, Louis Renault, who also lived in the French capital.

and one of his most memorable promotional initiatives was when he had the Eiffel Tower illuminated at night with the company's name and double-chevron badge. All over France, road signs bearing the message GIFT OF CITROËN were put up directing visitors to well-known landmarks.

André Citroën wanted the French public to be aware of his name from the youngest possible age. So in 1925 he introduced the Citroënnette, a child's pedal car which was a scaled-down version of the current touring model. The idea was that a baby's first words would be 'Mummy', 'Daddy' and 'Citroën'. These toys proved to be extremely popular and no less than half a million examples were made over a ten-year period.

A less desirable aspect of the urbane, eloquent André Citroën's character was that he was an inveterate gambler and he

9

regrettably chose to bring the ways of the casino to the more ordered environment of the boardroom. This risky approach to business was coupled with a barely disguised dislike of financiers and one of Citroën's oft repeated aphorisms was: 'As soon as an idea is good the money is of no importance.'

In 1929 production at Quai de Javel reached a record 92,000 cars, which made Citroën by far France's largest motor manufacturer. That year the firm introduced its first six-cylinder model, the AC6 powered by a 2.4 litre side-valve engine.

The model lines continued to evolve but, as will have been apparent, these Citroëns were popular and reliable though rather dull cars. But at the start of the 1930s the company embarked on the design of a revolutionary new front-wheel-drive model which was destined to be one of the most significant designs of motoring history. It was André Citroën's last great gamble and one, alas, that he was destined to lose.

Above: *In 1925 André Citroën introduced the Citroënnette, a child's pedal car, to publicise his product. This is a later, electrically powered version.*

Below: *Races were arranged for Citroënnette owners at seaside resorts. Here a group of children is about to compete in such an event at Deauville.*

10

Arguably one of the best-looking European saloons of the 1930s, the elegant low Traction Avant, styled by Flaminio Bertoni, introduced the concept of front-wheel drive to the mass market. This is the first photograph of the 1.3 litre 7A, as announced on 18th April 1934.

FIRST IN FRONT

In 1934, when Citroën's fabled front-wheel-drive Traction Avant model made its appearance, practically all cars were pushed along by their rear wheels. By contrast, most cars today are front-wheel driven or, in other words, the rest of the world has now caught up with André Citroën!

The model also introduced a host of features which were not in themselves new but were brought together in one car for the first time. What Citroën described as a 'new concept in motoring' was so in advance of its contemporaries that it remained in production until 1957, no less than 23 years after it had first appeared.

The starting point of the new Citroën was a unitary construction front-wheel-drive prototype, built by the Budd Corporation in the United States, where it was seen by André Citroën on one of his many visits to that country. It was only in 1933, just over a year before the car's announcement, that work began in earnest. Citroën himself faced considerable hostility from

Maurice Brogly, his head of design, and there were no components carried over from existing Citroën models (thus adding to the cost of the new car), so he recruited a talented graduate engineer named André Lefebvre to oversee the project. Lefebvre had worked for Renault since 1931, and before that for the eccentric car-maker Gabriel Voisin. He became project manager and other members of the talented engineering team were Flaminio Bertoni, responsible for styling, Raoul Cuinet, in charge of body construction, Maurice Sainturat, who designed the engine, Jouffret and Alphonse Forceau, responsible for the transmission, and Maurice Julien, in charge of the suspension.

The work proceeded rapidly, with the first prototypes on the road by August 1933, and the model was presented to the press on 18th April 1934, just thirteen months after Lefebvre's appointment. On 3rd May the first customer took delivery of a Traction Avant, which, by 1938, had

11

The 7CV Traction Avant unitary saloon bodywork in production at Citroën's Paris factory (above). It has yet to receive its engine and gearbox. These are pictured below and also revealed are the wishbone independent front suspension, by torsion bars, and drive shafts to the front wheels. This car is a prototype 7 Sport, the mechanicals of which differed slightly from those of the production cars.

Right: *In the 1930s the practice of designing a car's bodywork to conform to aerodynamic considerations to ease its passage through the air, and so improve petrol consumption, was virtually unknown in Britain. It was, however, practised on the continent and here a model of a Traction Avant has been photographed after wind-tunnel testing. The Citroën company has applied such disciplines to the design of its cars ever since.*

replaced all earlier Citroën designs.

When the new car was announced the French magazine *Le Journal* had proclaimed that the Citroën was 'so new, so audacious .... so different from all that had been done in the past that it merits the description of sensational'. But what made it so revolutionary?

While most cars at this time retained a separate chassis, the Citroën featured unitary construction, which dispensed with such a frame, and Bertoni produced an elegant four-door saloon with a distinctive and handsome sloping radiator grille, greatly enhanced by the presence of the Citroën chevrons. These features combined to suggest movement, even when the car was stationary. The use of front-wheel drive allowed the traditional propeller shaft to be dispensed with, so making the new Citroën much lower and more stylish than its contemporaries. Adventurously, the styling even took account of aerodynamics to make the car more 'slippery' as it passed through the air.

When the Traction Avant made its debut in 1934, most mass-produced cars used cheap though inefficient side-valve engines but the Citroën's engine had overhead valves and also used wet cylinder liners, which permitted a simpler construction of block. André Citroën had even envisaged fitting the car with a type of automatic transmission but the unit was

Right: *Because the Traction Avant did not have a separate chassis, Citroën had to convince the public of the strength of the unitary one by filming a 7A plunging over a cliff. The company had undertaken a similar publicity stunt when it introduced pressed-steel bodywork in the 1920s.*

13

The Citroën company was keen to emphasise the Traction Avant's low lines, so it recruited the Polish Count Chopski, who was 7 feet (2.13 metres) tall, to be photographed alongside one. This is an example of the longer wheelbase 1.9 litre 11CV Familiale (family) model of 1935, which was produced concurrently with the saloon, with six rather than four side windows.

dogged by problems and it was discarded only in February 1934, a mere two months before the car's launch. The replacement manual three-speed unit was designed and built in two weeks!

Suspension was, progressively, independent at the front and with a dead axle at the rear and the medium was simple but efficient torsion bars. The Citroën was the first road car to use these Porsche-patented units. Braking was similarly advanced and was a Lockheed hydraulic system.

Initially the Traction Avant was announced with two engine capacities. There was the basic 1303 cc 7A while the 7B had a 1529 cc engine and was also offered in open and closed two-door cabriolet forms. In addition there was the more powerful 7S, for Sport, with a 1911 cc engine, although in November 1934 this was renamed the 11 Légère (Light 11), which was destined to be the most popular Traction Avant of all. Later, in 1935, came the 7C, which represented a

combination of A and S concepts, and this 1628 cc model replaced the 7B at the bottom of the range.

The new range of front-wheel-drive Citroëns was displayed at the 1934 Paris Motor Show and this included three examples of the legendary 22CV, which, unfortunately, never entered production. It was powered by a 3.8 litre V8 engine consisting of two 11CV blocks mounted on a common crankcase. This Traction Avant would have had a top speed approaching the 90 mph (145 km/h) mark. It is believed that around twenty prototypes were produced though none survives.

All these Citroëns were technically the most advanced cars in the world at the time of their announcement but, tragically, at the end of 1934, André Citroën was in deep financial trouble. The cost of developing the new car, which also involved the partial rebuilding of the Quai de Javel factory, had placed an intolerable burden

14

The Traction Avant was also produced in two-door coupé form from 1934 until 1938. This is a 7CV car. The model's handsome radiator, incorporating the familiar Citroën double-chevron badge, greatly enhanced its appearance.

on the firm's never stable finances. Citroën appealed to the French government for help but it refused and suggested that he approach rival car-makers — his competitors. Fortunately he was spared that indignity and SA André Citroën was taken over by Michelin, its largest creditor, which had supplied the firm with wheels and tyres from the outset. André Citroën, who by this time had developed cancer, was broken-hearted and died on 30th July 1935, aged only 57.

In view of the speed in which the car had been conceived, the first examples did acquire a reputation for unreliability and this particularly related to the constant velocity joints, which not only had to steer but also convey power to the front wheels. These were soon replaced by bulky but effective Hardy Spicer universals mounted back to back.

However, its front-wheel drive endowed it with impressive road-holding and excellent cornering, regardless of weather conditions. In Britain *Motor Sport* magazine described it as 'one of the most controllable cars we know'. Perhaps the only quality lacking was performance and the

1930s elegance is well exemplified by this 11CV cabriolet version of the Traction Avant, produced in small numbers between 1934 and 1939. In Britain this model would be called a convertible.

This was a Traction Avant model which did not reach production. The 22CV version was to be powered by a 3.8 litre V8 engine and this cabriolet model was displayed at the 1934 Paris Motor Show. Its unique power unit can just be seen in the right background. Unfortunately Citroën's bankruptcy prevented the model from entering production.

smaller engines gave their models a maximum speed approaching 65 mph (105 km/h). The larger-capacity cars were rather faster and, after the war, an 11 could be taken up to 75 mph (121 km/h).

The Traction Avant was built not only in France but also at the firm's factory at Forest, Belgium, and in larger numbers at its British plant. These Slough-built cars differed from their French counterparts, apart from the inevitable right-hand rather than left-hand drive, in having a V-shaped bumper, 12 instead of 6 volt electrics, and leather upholstery replacing the rather sombre cloth of the French cars. Later the interior also benefited from the introduction of a wooden dashboard.

In 1938 the four-cylinder Traction Avants were joined by a six-cylinder version. The 15-6G was powered by a 2.8 litre engine which shared liners with the 11. It was, naturally, faster than its four-cylinder brothers and was capable of around 75 to 80 mph (120-8 km/h). In 1947 the model was redesignated the 15-5D. The new suffix stood for *droit*, which means 'right' in French, indicating that the engine turned clockwise though it had originally, and curiously, run anti-clockwise, hence the G for *gauche* in the title, meaning 'left'.

The outbreak of the Second World War in 1939 inevitably interrupted production, which ceased in France in November 1941 though some cars were built in Slough in

1942. Postwar manufacture concentrated on the 11 and 15 though until 1954 they were available only in black, like the Model T Ford. It was in that year that the 6 became the 15-6H. The new letter stood for *hydropneumatique*, for it was given a new type of rear suspension with pressure being maintained by an engine-driven pump. The reason for its fitment became apparent in October 1955, when at that year's Paris Show Citroën unveiled its new DS 19 model. It featured this new suspension and attracted the same sensational response that had greeted the Traction Avant 21 years previously.

That car remained in production for a further nineteen months. During its manufacturing life the low, front-wheel drive Citroën had become, like Camembert cheese or Gauloises cigarettes, an integral part of French life. During the war it was used by the German army and the Resistance alike and in peacetime it provided transport for gangsters and their police pursuers. In Britain the car was brought to a television audience by the exploits of Georges Simenon's Inspector Maigret. But the end finally came at Quai de Javel on 18th July 1957, when the last example of this pioneering front-wheel-drive model was built. In the 23 years since its announcement, a total of 759,123 had been produced, a demonstrable vindication, if one had been needed, of André Citroën and his 'new concept in motoring'.

16

The 1938 Traction Avants were joined that year by the 15CV model, powered by a 2.8 litre six-cylinder engine.   The entire range was greatly enhanced by the fitment of Michelin's handsome flat-spoked Pilote wheels and tyres, though they did not survive the war.

The last Traction Avant was built at Quai de Javel on 18th July 1957, 23 years after the model's introduction.   An 11D car, it was bought by M Dufur, a Citroën dealer from St Malo.

A prototype 2CV built in 1939. The corrugated front and single headlamp, fortunately, did not survive into the postwar years although the all-independent suspension, hinging front windows and roll-back roof were all used on the production cars.

SPARTAN SURVIVOR

As will have been apparent, the Traction Avant was a modern, stylish and revolutionary concept. But the two-cylinder 2CV, or Deux Chevaux ('Two Horsepower'), which Citroën announced in 1948, was as distinctive in concept as its illustrious predecessor though where that car had been elegant the 2CV was functional, utilitarian and, some would say, downright ugly. It was like no other car on the road and over seven million were built, making it numerically the most successful Citroën ever. The 2CV was in production for over 41 years with manufacture ceasing in 1990.

It was not an engineer but Michelin's architect, Pierre-Jules Boulanger, who can be considered to be the 'father' of the 2CV. He had first become acquainted with the Michelin family in 1905 and after the First World War had been responsible for designing homes for the company's workforce. It was early in 1936, which was the year after the company's

takeover of Citroën, that Boulanger, backed by Pierre Michelin, approached the chief engineer, Maurice Brogly, and asked him to design 'an umbrella on four wheels'.

Boulanger was thinking in terms of a cheap car that would appeal to France's rural community and to those thousands of people who had never owned a car. In view of this, he maintained that the projected car should be able to carry two peasants in their working clothes, together with 110 pounds (50 kg) of potatoes and have a top speed of 37.5 mph (60 km/h) Petrol consumption would be accordingly modest and the car would consume a mere 3 litres of petrol per 100 kilometres, which was over 90 miles per gallon. A key element of this specification was that it should be drivable over indifferent roads and rural tracks. To illustrate his point, Boulanger maintained that, if the occupants chose to place a carton of eggs on the rear seat and the car was then driven

18

over a ploughed field, not one egg would be broken.

Such a concept did not appeal to the conservative Brogly, so responsibility for the project was given to André Lefebvre, who had turned Citroën's dream of the Traction Avant into a reality. He recruited some of the key engineers who had been involved in that model's creation to work on this new small car, which was also to have front-wheel drive. Styling, if that is the correct word, was again assigned to Flaminio Bertoni; young Jean Mauratet would engineer the bodywork structure; transmission was Alphonse Forceau's established brief, while Marcel Chinon, formerly chief engineer of the Amilcar company, was responsible for overall co-ordination of the project.

Work was soon under way on the car which was designated the TPV, for *très petite voiture* (very small car), and a mock-up was built at Citroën's experimental workshop in the Rue du Théâtre, Paris. There it was viewed by Boulanger, who was a tall man, and it is said that the height of the car's door line was dictated by him being able to enter the driving seat with ease, wearing a hat!

The first prototype TPV was on the road at the beginning of 1937 and by the end of the year there were some twenty of these experimental cars in existence. They possessed the same ungainly functional lines as the production models of eleven years later, though they were more curvilinear than angular in profile: high four-door saloons with canvas roll-back roofs and fitted with hammock-like seats. A distinctive feature of the design was that the front of the car was not only made of aluminium but was corrugated, like the fuselage of a Junkers aeroplane, for strength. Only one headlamp, the legal minimum, was fitted. At this stage there was no engine, so a 500 cc BMW motorcycle unit which shared the two-cylinder horizontally opposed configuration of the finished product was initially employed. This was dispensed with as soon as the Citroën-designed water-cooled 375 cc engine, the work of Maurice Sainturat, was completed.

Pierre Boulanger (1885-1950), 'father' of the 2CV. He was made Citroën's managing director in 1935 and, following the death of Pierre Michelin in 1937, became director general in 1938; his tenure was cut short by his own untimely death in a road accident.

Lefebvre had been intent on saving weight and had built the car around an aluminium platform chassis with all-independent suspension (to preserve those eggs!) by torsion bars which had proved their worth on the Traction Avant. The rack and pinion steering and hydraulic brakes also followed precedent.

By August 1938 Pierre Boulanger felt that the concept was sufficiently developed for production proper to begin in May 1939, with a launch in the autumn of that year. There were, however, delays. The extensive use of aluminium proved to be a major problem and the material was found to be particularly difficult to spot-weld, so it was not until 2nd September 1939 that the first of a batch of 250 cars was built. The following day France, along with Britain, declared war on Germany and the Second World War had begun.

19

Interior of the 1939 prototype with rudimentary hammock seats, dashboard-mounted gear lever and hand-operated windscreen wiper. This car is one of three surviving prototypes.

The production of the 2CV in its original form was, perhaps fortunately, interrupted by the outbreak of hostilities and the German occupation of France the following year. Nevertheless, development work continued. Boulanger ordered the destruction of the surviving 249 incomplete 2CVs and a further radical change to the design was made in 1944 when Walter Becchia, who came to Citroën from Talbot in 1941, suggested that, in view of the difficulties in starting that had plagued the original engine, its water cooling should be replaced by air, which also had the virtue of being a cheaper, if noisier, solution. He redesigned Sainturat's unit while still retaining its two-cylinder overhead-valve configuration and 375 cc capacity. Becchia also produced a new gearbox with four speeds and that was something which even the Traction Avant did not possess!

The 2CV was finally unveiled at the Paris Motor Show of 1948, after a lengthy,

The President of France, Vincent Auriol, visiting the Citroën stand at the 1948 Paris Motor Show, where the 2CV was launched. He can be seen next to the white-haired Pierre Boulanger inspecting the middle car.

A driver and three passengers enjoying the delights of 2CV motoring. This is a 1950 car. The double chevrons on the radiator grille lost their oval surround in the 1955 model year.

twelve-year gestation. It was not received with universal acclaim. Some members of the French press criticised the new Citroën though *The Autocar* from Britain was more charitable, declaring that it was 'the work of a designer who has kissed the lash of austerity with almost masochistic fervour ... but it is full of original ideas planned to cut back weight and cost'.

At the beginning the 2CV sold for 225,000 francs, which was the equivalent of £212. The car could be clearly seen to have evolved from the prewar prototypes though a more conventional front replaced the corrugated ripples of the original. There were more significant changes below the surface. The aluminium parts had given way during the war to pressed

The interior of a 1953 2CV with new tartan upholstery. This picture clearly displays the space-saving attributes of front-wheel drive, there being no obstructive transmission tunnel and therefore a desirably flat floor.

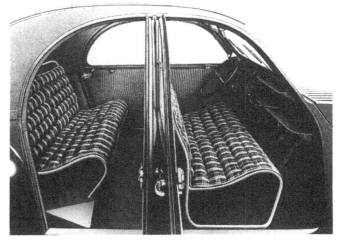

A happy French family pictured in a 2CV against the background of Mont St Michel. This is the AZL version introduced in 1957, identifiable by the trim strip in the centre of the bonnet.

sheet steel and, after a number of changes, the design of the interconnected all-independent suspension was finalised. Each wheel of the front leading and rear trailing arms now possessed its own enclosed spring damper.

Performance was modest: the new car could achieve around 40 mph (64 km/h), though this has to be offset against an impressive fuel consumption of 55 miles per gallon, or 19 km per litre. Acceleration was poor but that did not matter. The 2CV was a willing workhorse that could carry four people and therefore represented family transport. It was just the type of car to find favour in a Europe that was being rebuilt from the ashes of conflict.

Such was the demand for the 2CV that even Citroën was taken by surprise and in 1950 delivery times ran to no less than six years. Production therefore began in Belgium in 1952 and, in the following year, at Citroën's Slough factory. But the British did not take the 2CV to their hearts

in the 1950s and a lack of demand in Britain forced Slough to discontinue assembly in 1959. It tried again that year with the Bijou, which retained the 2CV's mechanicals but was fitted with an updated and DS-related two-door glass-fibre body, the work of Peter Kirwan-Taylor, who had been responsible for the superlative lines of the Lotus Elite of 1956. However, there were few takers and only 213 Bijous were produced before Slough quietly dropped the model in the spring of 1964.

Changes were, in the meantime, being made to the 2CV itself. The most significant occurred in 1954 when its capacity was increased to 425 cc; this enabled the top speed to approach the 50 mph (80 km/h) mark. In 1957 annual production of the 2CV exceeded 100,000 for the first time with 107,251 examples built, and output for this basic model peaked in 1968 when 168,384 cars were produced. By this time the original 2CV had been joined,

22

Above: *Improvements to the 2CV for the 1961 model year included a new bonnet and radiator grille, and this AZAM model, which appeared in 1963, is distinguished by chrome hub caps, tubular overriders and trim strips around the windscreen and side windows.*

Below: *The 2CV's roll-back roof is here being put to good use. The plastic grille, rectangular headlights and black bumper inserts appeared for the 1975 season.*

in 1961, by the Ami, a rebodied and better-equipped 602 cc version, which remained in production until 1978.

By the mid 1960s the 2CV, with its roots firmly planted in the 1930s, had been in production for seventeen years. It seems more than likely that the Dyane, launched at the 1967 Amsterdam Motor Show, was conceived as a replacement though the older model was destined to outlive it. The new car again used the 2CV chassis and 425 cc engine though the fitment of the Ami's final drive made the model a 60 mph (96 km/h) car. The new body was essentially an updated version of the 2CV one, and there was a practical and increasingly fashionable tailgate.

In 1968 the Dyane was offered with the Ami's 602 cc engine, which made speeds approaching 70 mph (113 km/h) theoreti-

23

The 2CV acquired its additional rear side windows in 1966 though the Charleston originated as a limited-edition model in 1981. Because of its popularity, it became part of the range. This black and maroon colour scheme echoed the sweep-panel treatment of some American and European cars of the 1920s.

cally possible. Production was on a par with the 2CV, though greater than that of the Ami, and the Dyane continued to be built until 1982. Its chassis also formed the basis of the plastic-bodied Jeep-like Mehari, named after a dromedary favoured by desert tribesmen, which arrived in 1968. A four-wheel-drive version of this long-running model was developed for the French army in 1981.

In the 1970s the 2CV proper enjoyed something of a revival and between 1972 and 1979 annual output never dropped below the 100,000 mark. From 1968 the cars had become more colourful with the original drab greys and later fawns being replaced by more cheerful reds and blues. A major factor in this renaissance was the outbreak, in September 1973, of the Arab-Israeli war, which brought a steep rise in oil prices, and customers turned again to economical cars that were cheap to run. In 1974 the 2CV was reintroduced in Britain. Not only did it become popular as a second car but it also gained a faithful following amongst the younger genera-

The 2CV finally ceased production, after 42 years, on 27th July 1990, the last one being turned out at Citroën's factory at Mangualde, Portugal.

The 2CV-related Ami-6 saloon arrived in 1961. Citroën claimed it to be 'the world's most comfortable medium-sized car' though the inclined rear window, in the manner of Ford's 105E Anglia of 1959, was a controversial feature.

tion. There was no better indication of the model's enhanced status than when James Bond drove one in the film *For Your Eyes Only*, released in 1981.

But the 2CV was nearing the end of its production life. Sales began to dwindle in the late 1980s and the car would have been unable to comply with future European safety and emissions regulations. When Citroën made plans to rebuild part of its by then ageing Paris factory, production was transferred in February 1988 to its plant in Mangualde, Portugal. In 1989 output dropped to a mere 19,000 cars, with Germany as the principal customer, having bought 7866, compared with France's 5231 and Britain's 3200. At the end of July 1990 the final Deux Chevaux was built. During its production life the car had been assembled by Citroën at its plants as far afield as Argentina, Iran and Madagascar. Such had been the success of Pierre-Jules Boulanger's 'umbrella on four wheels' that it had found favour all over the world, even in countries where it seldom rained!

The Ami-8 replaced the 6 in 1969 though this is the outwardly similar Ami Super of 1973 which was powered by the air-cooled 1015 cc flat-four engine from the Citroën GS.

25

The DS, the Traction Avant's long-awaited successor, finally appeared at the 1955 Paris Motor Show. Just as advanced as its predecessor had been, it perpetuated the front-wheel-drive theme, coupling this with advanced, aerodynamic-influenced styling and hydropneumatic suspension.

STAYING IN FRONT

In 1955 Citroën unveiled the DS, the Traction Avant's successor, which was destined for a twenty-year production life. The range's increasing technical sophistication was reflected in the complex SM Grand Tourer of 1970. After a downturn in the world economy from 1973 the company got into financial trouble and the following year it was taken over by Peugeot. Since then Citroën has benefited from an essential element of rationalisation, the current range combining a dash of *élan* with the latest in automotive technology.

Inevitably it was a Paris Motor Show where Citroën unveiled the DS 19 and, as in the 1934 and 1948 events, in 1955 the company's stand was besieged by visitors anxious to see the new car. They were not disappointed. The futuristic DS, also masterminded by Lefebvre and styled by Bertoni, looked sensational with, in the interests of aerodynamic efficiency, its plunging bonnet line and absence of radiator grille, an approach which would not be embraced by the rest of the motor industry until the 1970s. The long-stroke 1.9 litre engine was the same as its predecessor's though it had an alloy hemispherical cylinder head. Otherwise everything else was new, notably the self-levelling all-independent suspension, which used oleopneumatic struts and was pressurised by Lockheed fluid from an engine-driven pump, which also controlled the steering and even the clutch for the semi-automatic gearbox. In addition it actuated the brakes and the DS used inboard discs at the front and thus had the distinction of being the first production road car in the world to be fitted with them.

With its streamlined shape, the DS 19 was able to transport its occupants, in considerable comfort, at speeds around 80 to 85 mph (129/137 km/h), while the road-holding and handling, because of the front-wheel drive, were by then formidable Citroën attributes.

In 1956 the supplementary ID 19, powered by the Traction's 1.9 litre engine, and with hydraulics confined to the suspension, was introduced. From 1966 the ID inherited the DS's engine, two new short-stroke units of 1985 and 2175 cc being fitted to the DS 19 and a new DS 21 respectively. The advanced body lines remained essentially unchanged though

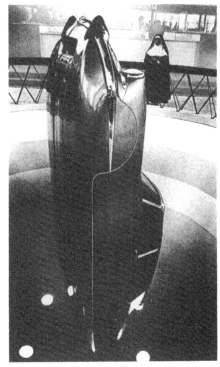

Right: A nun contemplates a dramatic presentation of the DS's hull. For aerodynamic considerations the radiator was relegated to below the bonnet line, an arrangement that the rest of the motor industry did not adopt until the 1970s.

Below: The DS's interior was as advanced as its outward appearance and the single-spoke steering wheel was a notable feature. This is the dashboard of the subsequent and better-equipped Pallas version of the DS 19, introduced in 1965.

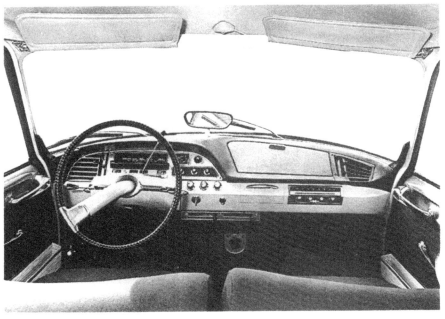

the range acquired faired-in headlamps, making the cars look even more shark-nosed, and the DS was fitted with swivelling headlamps. The simpler ID line was discontinued in 1969 and the 21 was fitted with fuel injection in 1971. It was on 24th April 1975, nearly twenty years since it had been announced, that the last of 1.4 million DSs was built. The model was also produced by Citroën's British factory at Slough, which assembled it for the ten years between 1956 and 1966. This was the last model to be built there. Thereafter cars were imported directly from France.

The DS's replacement, the CX, had entered production in 1974. A smaller car than the D Series, it used that model's proven hydropneumatic suspension and engine, which was now mounted transversely rather than in line. Available in 2, 2.2 and later 2.4 litre forms, the four-door saloon, with large opening tailgate, maintained the company's traditions for advanced styling and it remained in production until 1989.

In 1970 Citroën had introduced a completely new model to fill the gap between the 2CV and the DS. The 1 litre GS was a five-seater four-door saloon with distinctive cut-off tail and powered by a new air-cooled alloy flat four-cylinder engine with single overhead camshafts per cylinder bank. Like the DS it was fitted with all-independent hydropneumatic suspension and all-round pressurised disc brakes. It was Car of the Year in 1970 and, although top speed was over 85 mph (138 km/h), it was a little underpowered for many tastes and a 1.2 litre version was added for 1973. The millionth GS was built in 1977 and the model endured until 1985 after a respectable fifteen-year production run.

In 1968 the Italian Fiat concern took, despite protests from the French government, a 15 per cent stake in Citroën but earlier in the same year the French firm had purchased a majority shareholding in the Italian Maserati company, which it consolidated by outright ownership in 1971. Citroën now had a suitable engine for its ambitious and extravagant Grand Touring project: the SM of 1970. It was a car, one feels, of which André Citroën would have approved! Based on the DS's

For 1968 the DS's headlamps were enclosed for greater aerodynamic efficiency. On the Pallas versions of the DS 19 and DS 23 the inner quartz iodine lights were connected to the steering and swivelled accordingly. This is a DS 23 with electronic fuel injection.

Citroën's Grand Tourer, the SM of 1970, was DS-based. Powered by a 2.7 litre V6 Maserati engine, only 12,920 were built before production ceased in 1975.

complex mechanicals, its 2.7 litre V6 engine was related to the Maserati's V8 Indy unit. Tailored to Citroën's front-wheel-drive requirements, it was the fastest, at 137 mph (220 km/h) and also the most expensive Citroën ever, selling for 51,800 francs in France and £5480 in Britain. Regrettably the car failed to attract sufficient buyers: only 12,920 were built,

and Citroën's problems were compounded by the 1973 global oil price rise. The liaison with Fiat was unscrambled that year and, in December 1974, Michelin sold Citroën to Peugeot, which quickly disposed of Maserati.

The economical Visa of 1977 reflected some important cross-pollination and rationalisation between the two concerns.

The DS was replaced in 1974 by the CX. This is the top of the range 25 GTI Turbo of 1984 with a top speed of over 125 mph (201 km/h). Like its predecessor, the CX enjoyed a long manufacturing run and was discontinued, after fifteen years production, in 1989.

The immensely popular BX five-door hatchback, introduced in 1982, offered traditional Citroën hydropneumatic suspension and a choice of Peugeot engines from 1.4 to 1.9 litres capacity.

It consisted of the hull of the Peugeot 104 united with Citroën's 602 cc flat twin engine and, from 1979, it was available with a 1124 cc Peugeot unit.

A new generation of models appeared in the 1980s. The middle-sized BX five-door hatchback powered by 1.4 and 1.9 litre Peugeot engines, though with the traditional hydropneumatic suspension, was introduced in 1983 and became a best-seller. Less sophisticated, however, was

the smaller, robot-built AX of 1986, which was offered in three engine sizes of 954, 1124 and 1360 cc.

The flagship of the range, and the CX's replacement, arrived in 1989. The XM was a five-door luxury saloon, styled by the Italian Bertone company, with computer-controlled hydropneumatic suspension which resulted in the most advanced system ever fitted to a production car. It was available with

Citroën's supermini, the AX of 1986, was a light, economical robot-built hatchback that was offered with a new generation of engines ranging from 954 to 1360 cc. It survived until 1997.

The Bertone-styled five-door XM hatchback was the top of the range. Introduced in 1989, it could be specified with a variety of engines, from 2 and 3 litre petrol to 2 litre unblown and turbocharged diesels. A sophisticated electronic control system adjusted the car's suspension in relation to its speed and cornering with the road surface. Production ceased in 2000.

engines from 2 to 3 litres in size. Like its predecessor, the Citroën XM was hailed as a European Car of the Year, displaying all the flair and engineering excellence that have made the highly individual products of this remarkable French company so famous.

It survived until 2000 and during the previous decade Citroën had continued to produce distinctive, sometimes quirky models, whilst also maintaining its tradition for sophisticated engineering and suspension.

In 1993 came the mid-range Xantia, enhanced with Bertone's crisp five-door hatchback styling, and a range of petrol engines from 1.6 to 2 litres, a 1.9 litre diesel and turbo diesel. Offered with Hydractive self-levelling suspension, the name of this BX replacement was meaningless but was reckoned to be classy...

Citroën entered the people-carrier market in 1994 with the Synergie, developed in conjunction with the Fiat Group. Also badged in-house as the Peugeot 806 and as the Fiat Ulysse and Lancia Z, it was built at a new factory at Valenciennes geared to produce 130,000 examples a year. It was powered by a choice of 2 litre, 2 litre turbo and 1.9 litre diesel engines, and, whilst accommodation was good (up to eight could be carried), its bland styling was hardly in the Citroën tradition. It survived until 2002, when it was replaced by the better-looking C8, which sprang from the same corporate parentage.

Much more successful was the compact multi-purpose Xsara Picasso for 1999. It was based on the Xsara, which replaced the ZX in 1997 and was available in five-door hatchback or coupe forms with a variety of 1.4 to 2 litre petrol and 1.9 litre diesel engines. Destined to provide Citroën with a strong seller, its Picasso derivative had individual seating for five in, above all, a body that showed flair and was functional. This, coupled with an inspired television advertising campaign that featured recalcitrant robots, lead to the Picasso becoming, in 2002, Britain's top-selling multi-purpose vehicle (MPV).

The Xantia was replaced for 2001 by the C5, in profile a chunky four-door saloon but in fact a hatchback. It featured the latest Hydractive 3 suspension, which is an electro-hydraulic and self-levelling interconnected system, the origins of which can be traced back to the Traction Avant 15-6H of 1954. It comes into its own on very rough roads. Roomy and refined, with engines ranging from 1.8 to 3 litres in size, the latter version is capable of over 140 mph (237 km/h).

Citroën's supermini had been the 1/1.6 litre Saxo, which replaced the AX in 1996, and it was also badged a Peugeot 106. In 2002 this gave way to the C3, based on the platform of the projected Peugeot 207, with curvy lines and available in 1.4 and 1.6 litre guises. For 2003 came the C3 Pluriel convertible, which injects an element of fun to the range. A company that produced the 2CV has, creditably, never taken itself too seriously!

Citroën's best-selling Xsara Picasso is available with a range of power units and trim options. This version can be powered by a choice of 1.6 or 1.8SX petrol engines or a 2 litre SX diesel unit.

FURTHER READING

The following list contains some older books that may be found in second-hand bookshops, autojumbles or libraries.

Bobbitt, Malcolm. *The British Citroën*. Transport Publishing Company, 1991.
Pagneux, Dominique. *Citroën un Genie d'Avance*. ETAI, 2001. French text.
Pressnell, John. *Citroën Traction Avant*. The Crowood Press, 2005.
Reynolds, John. *André Citroën: The Man and the Motor Cars*. Sutton Publishing, 1996.
Reynolds, John. *The Citroën 2CV*. Sutton Publishing, 1997 and 2001.
Reynolds, John. *Citroën Daring to Be Different*. Haynes, 2004.
Reynolds John. *Citroën from A to X*. Citroexpert, 1998.

JOURNALS
There is no magazine that deals exclusively with aspects of Citroën history but the following carry articles on the make from time to time.
The Automobile, Enthusiast Publishing Ltd, PO Box 153, Cranleigh, Surrey GU6 8ZL.
  Telephone: 01483 268802. Website: www.the-automobile.co.uk
Classic Cars, Bauer, Lynchwood, Peterborough Business Park, Peterborough PE2 6EA.
  Website: www.classiccarsmagazine.com
Classic & Sportscar, Teddington Studios, Broom Road, Teddington, Middlesex TW11 9BE.
  Website: www.classicandsportscar.com

CLUBS
Citroën Car Club, PO Box 348, Steyning, West Sussex BN44 3XN.
  Telephone: 07000 248258. Website: www.citroencarclub.org.uk
Deux Cheveux Club of Great Britain, 2CVGB Membership Secretary, 19 Railway Road, Wisbech, Cambs
  PE13 2QA. Website: www.2cvgb.co.uk
Traction Owners' Club, Membership Secretary: John Oates, 55 The Knoll, Tansley, Matlock, Derbyshire
  DE4 5FP. Website: www.traction-owners.co.uk

PLACES TO VISIT

Museum displays may be altered and readers are advised to check before travelling that relevant items are on show and to ascertain the opening times.

GREAT BRITAIN
Haynes Motor Museum, Sparkford, Yeovil, Somerset BA22 7LH.
  Telephone: 01963 440804. Website: www.haynesmotormuseum.co.uk
National Motor Museum, Beaulieu, Brockenhurst, Hampshire SO42 7ZN.
  Telephone: 01590 612345. Website: www.beaulieu.co.uk

FRANCE
Henri Malartre Motor Museum, 645 Ruc de Musee, 69270 Rochetaillee-sur-Saone.
  Telephone: +33 (0) 478 221 880.
Musée National de l'Automobile (The Schlumpf Collection), 15 Rue de l'Epee, 68100 Mulhouse,
  Haut-Rhin. Telephone: +33 (0) 389 332 323.
Musée National des Techniques, Conservatoire National des Arts et Métiers, 292 rue Saint-Martin,
  F-75141, Paris.
Musée National de Château de Compiègne, place du Général-de-Gaulle, F-60200, Compiègne, Oise.
  Telephone: +33 (0) 344 384 702.

Printed and bound by CPI Group (UK) Ltd, Croydon, CR0 4YY

11/10/2024

01043558-0005